Nifty E-Z Guide to PSK31 Operation

One of the guides in the Nifty! Ham Accessories Easy Guide Series

www.niftyaccessories.com

Copyright

Disclaimer and Limitation of Liability

Contents

About This Guide

This guide is designed to help you get PSK31 up and running quickly and easily. After a short introduction to the PSK31 digital operating mode, subsequent chapters walk you through getting your transceiver interfaced to a home computer and downloading and using the *DigiPan* free-ware program, which is relatively simple and can be made to work with just about any HF bands capable ham radio.

We will go light on theory, concentrating on the practical issues of getting things connected and making PSK31 contacts.

Three chapters are devoted to different methods of interfacing your transceiver to a computer. These include a simple build-it-yourself interface and two different commercially available interfaces. Reviewing these three chapters will give you a good idea of how you want to approach your own installation.

After selecting an interfacing method and getting things connected we progress to downloading, installing and configuring the *DigiPan* software.

Once everything is configured, we explore how to find and receive PSK31 signals and cover the fine points of adjusting your transmitter for proper operation.

Let's get started!

Chapter 1: **PSK31**

PSK31 is a popular mode of digital mode of operation that allows radio amateurs to easily communicate via text messages using a standard personal computer, a PSK31 software program, and a SSB transceiver. The trick that makes all this work is some very clever software that configures the PC's sound card to operate as a digital modem.

History

In 1998 Peter Martinez, G3PLX, developed the PSK31 digital encoding method, *Varicode*, and the PC software necessary to make it work. His system used the PC's sound card to send and receive phase-shift keying transmissions using ordinary amateur HF transceivers. Due to the code's robustness and its narrow 31 Hz bandwidth, PSK31 has the amazing ability to provide communications in spite of noisy band conditions and poor propagation.

Since Peter's early efforts, many others have contributed their expertise by developing ever more capable and easy-to-operate software. The combination of readily available software, low-cost PC's and superior operating performance has led to PSK31 being adopted as one of the primary modes of amateur digital communication.

PSK31 Coding

Instead of using the familiar AFSK *Audio Frequency-Shift Keying,* technique, commonly used for RTTY communication, PSK31 uses *Phase-Shift Keying*. In AFSK two different audio tones are used to represent digital data transitions. However, in PSK the **phase** of a sound card generated 31 Hz audio tone is shifted 180 degrees (a phase reversal) to generate digital signal transitions. This phase reversal technique is referred to as *Binary Phase-Shift Keying* or BPSK. Phase shift modulation is much less affected to fading and noise than AM modulation, which is one of the reasons PSK works so well.

The bandwidth requirements for digital communication using AFSK is related to the communication rate (bits per second) and the amount of audio frequency shift employed, typically on the order of hundreds of Hertz. But since PSK uses a *Phase Shift* technique, the frequency remains the same and it therefore occupies only a very narrow bandwidth determined by the data bit rate.

The “31” in PSK31 refers to the data bit rate, which is the same as the bandwidth needed – a very narrow 31 Hz. While 31 bits per second can readily keep up with the average typist, its relatively low speed, and lack of error correction methods, makes it unsuitable for transmitting large data files or for transmitting data that must be absolutely correct.

Similar to Morse code, PSK31 uses a variable number of digital bits (ones and zeros) to represent different alphanumeric characters. Variable length PSK31 codes are inherently more efficient for serial data communication than Baudot or ASCII, which use *fixed length* 5 bit and 7 bit codes. In PSK31 character codes range from 1bit (which represents a space) to codes 10 bits in length (which are used in various combinations for punctuation and special symbols). Shorter codes are assigned to the most commonly used characters such as “a”, “e” etc., significantly improving serial data throughput for narrow bandwidth communication applications. G3PLX referred to his variable length coding system as *Varicode*.

PSK31 Code Variants

Like most things that have generated a lot of interest, a number of variations to the basic BPSK coding technique have been developed. Probably the foremost of these variations is *Quaternary Phase-Shift Keying*, QPSK31, which provides significant error correction capabilities. Operating PSK in the QPSK mode provides nearly 100% copy under most conditions.

Another PSK31 variant gaining popularity is PSK63, especially with contesters because of its 100 wpm character transmission rate. It is essentially the same as PSK31, only faster and requiring a 63 Hz bandwidth. PSK63 mode comes in several flavors: BPSK63, which is essentially a faster variant of BPSK31, QPSK63 with error correction, and PSK63F with Forward Error Correction.

Despite the above "improved" variants, most of the action on the bands is still with the original PSK31 BPSK mode of operation. There are many other variants out there, but most are experimental.

Resistance to Interference

With a bandwidth requirement of only 31 Hz, PSK31 provides a significant advantage when the bands are noisy and crowded. For those familiar with CW operation, the signal reception benefits of being able to filter down to a narrow bandwidth are well understood.

In addition to the advantage of an extremely narrow bandwidth, the cleverly designed PSK31 decoding software has a number of innovative techniques for improving reception. The software has been optimized to look for the expected 180-degree BPSK phase-reversal transitions and uses that fact to stay in sync with the transmitting station. If the signal is momentarily disrupted, the software can quickly re-synchronize itself in just a few characters.

Experience has shown that PSK31 can often provide useful communications under poor propagation conditions and interference, which may cause voice and other digital modes to fail. Under adverse signal conditions it is at least as good as CW and significantly better than RTTY.

Because of the mode's almost miraculous ability to "pick a signal out-of-the noise", even under adverse band conditions, those who operate at low-power levels and lack large antennas especially appreciate the advantages of using this mode of operation.

With the software that is now available, PSK31 is as easy, or easier to use as the older RTTY method of communication and has the added advantage of being much more robust under weak-signal conditions. And when the bands become crowded during contests, PSK31's narrow bandwidth has quite an advantage over other wider modes of communication.

PC to Transceiver Interconnect

Fundamentally the required computer to radio interface is quite simple. Since the sound card is used to generate the audio signal that modulates the transceiver, it follows that the sound card's Speaker Out jack (or preferably if your system has one – the Line Out jack) is connected to the MIC input of the transceiver. Likewise the transceiver's Speaker Output is connected to the sound card's Microphone Input (or Line In if it has one) so that received signals can be decoded.

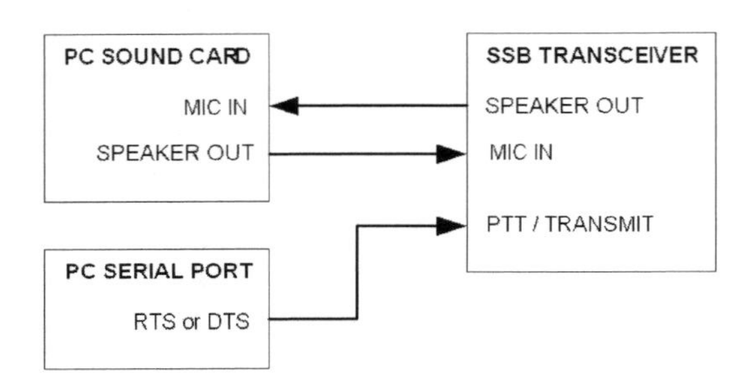

Simplified PC to Transceiver Interconnect Diagram

As shown by the above diagram, the basic interconnect concept is quite straightforward. But as we shall see in following chapters, because of signal level differences, potential ground loop problems, sound card volume setting issues and the desire to use a transceiver's auxiliary interface connectors, getting things connected and working properly can become a bit more complex than the above diagram implies.

PSK31 Software

Fortunately, today we are blessed with a large number of PSK31 programs to choose from. Some can be downloaded free from the Internet; others are commercial programs that support logging and a variety of modes of operation. Regardless, most of these programs are quite good and support a variety of features beyond basic PSK31 operation.

For purposes of this guide we will be using *DigiPan*, primarily because it will work with just about any ham transceiver, is relatively simple to use, its free and has been extremely popular with hams for many years. *DigiPan*, which stands for Digital Panoramic Tuning, is most noted for its ease of tuning using its *waterfall display* and the ability to detect and decode multiple transmissions at once. The program is designed so that most operations, including tuning to transmitting stations, are accomplished by a simple click of a mouse. Downloading, setting up and using *DigiPan* is covered in Chapter 3.

In addition to *DigiPan,* many of the other PSK31 programs also have *waterfall displays* and similar tuning capabilities. The latest version of *DigiPan* is 2.0, which was released in October of 2006. *DigiPan* and other early digital modes programs were the source of inspiration for software developers and there are a lot of other PSK31 programs out there. Some programs are offered as freeware and others may be purchased. Many of these programs are quite outstanding. Refer to Chapter 8, "Onward to Other Digital Modes" for a listing of other digital mode programs and URL's where you can find more information and download the programs.

Chapter 2: **Homebrew Radio Interface**

If you enjoy putting things together from scratch, making your own PSK31 radio-to-PC interface may be the way to go. On the other hand, if you prefer to buy a hardware interface solution, review Chapter 6, "Rascal GLX Radio Interface" or Chapter 7, "SignaLink USB Radio Interface" instead.

The interface presented in this chapter is relatively simple to build and will work reasonably well in many cases. While there may be some operational disadvantages over some of the better commercial solutions, this lets you tryout PSK31 operation at the lowest possible cost. A lot of hams use this method and depending upon how many of the parts you can find in your shack, it might be done for no money out of pocket.

Connecting the Sound Card to the Transceiver

On the sound card side, things are straightforward, if your PC has Line-In and Line-Out jacks you should use them as your first choice. If, as on most laptops, you only have Microphone-In and Speaker-Out, use those instead.

The advantage of using your sound card's Line-In and Line-Out jacks is that they have separately settable volume levels and will not interfere with the volume levels you may be using for a headset or an external set of speakers. From a cable fabrication standpoint, it probably doesn't make any difference, as the jacks on your PC should be the same. Typically these are 3.5mm (1/8") audio style jacks.

Use Shielded Audio Cable

On the subject of radio-to-PC interface cables, to minimize interference from stray RF and other signals in the shack, it's best to use shielded audio cables for the interface. These can be made from "surplus" microphone or speaker cables that you might have lying around. This is especially convenient since they may come equipped with the right sized jack for plugging into your PC. In general, it's best to keep these cables relatively short. The longer they are, the more susceptible they become to picking up RF or other interference.

In addition to using shielded cable, to further protect from possible signal interference pickup, it's a good idea to place a ferrite core around both the Mic/Line-in and Spkr/Line-out cables at the end where they plug into the PC.

Radio-to-PC Cable Interface Schematic

The schematic below shows how your interface cables should be wired. The two resistors shown act as a 100:1 voltage divider to reduce the relatively strong speaker-out signals from the PC to make them compatible with the low-level microphone input on your radio.

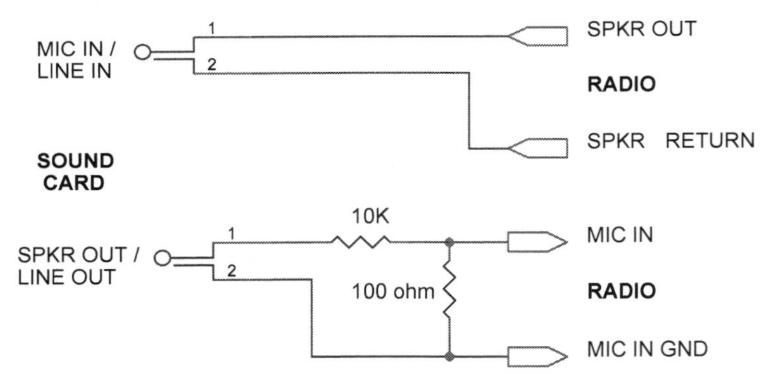

Depending upon your radio, you may be able to connect to an auxiliary accessory socket at the rear of the radio rather than connecting directly to the speaker and microphone inputs. Check the documentation in your radio's manual to see if this is possible. Your radio may have come with the required accessory connector in the bag of miscellaneous parts that originally came with the radio.

Just Receiving PSK31 to Start

If you are in a hurry to get started receiving PSK31 to see how things work, you could get by with just making the top cable in the above diagram. This will allow you to route received signals to the PC so that you can monitor band activity and get used to using your radio and the PSK31 software without having to worry about transmitting.

Keying your Transceiver

In addition to a cable for routing signals between the PC and your radio, you need to decide how you are going to key your radio's transmitter. To start with you could just manually key your transmitter, much as you would using push-to-talk on your microphone. Depending upon your radio, you might be able to flip a switch on the front panel or perhaps modify an old microphone to disable the sound pick up element but allow PTT keying.

While manually keying the transmitter may work for initial testing, this will soon get tedious and you will ultimately want the software to take care of keying transmission. For most of the PSK31 software programs this is typically done by connecting a signal on your PC's serial port to the radios' transmit keying signal input.

Depending upon the type of PC you have this may present a problem. Older PC's and laptops almost always come with a serial port that can be used. More recent vintage PC's however generally eliminate the serial port in favor of one or more USB ports.

If your PC or laptop does not have a serial port you could use a USB to serial port adapter if you have one. If you don't have an adapter you could buy one, but if you are going to have to spend some money on that, it might be time to consider purchasing a commercial PSK31 hardware interface like the TigerTronics, *SignaLink USB* discussed in Chapter 7.

Using VOX to Key the Transceiver

If you don't have a serial port available, or your radio lacks a good way of being externally keyed, using VOX is another option that might work. On many radios, VOX only works on the microphone input connector and not on the rear panel accessory connectors, which are usually geared towards TNC interfacing. If you can get VOX to work, it would save building the following transmitter keying circuit.
There are two potential problems with using VOX that you need to keep in mind, and look for if you try this approach. One is the VOX keying delay—the time it takes for the VOX circuit to detect the

audio input signal and then actually start transmitting. If this time is excessive, you might have the first characters of your transmissions clipped. The second problem is that your radio may inadvertently start transmitting whenever your computer produces any audio: the occasional beeps that are generated for all kinds of reasons, and of course other audio applications that produce sound: music, arriving email, video clips, etc. Not the kind of thing you want to happen in the PSK sub-band, or anywhere else for that matter; music and other un-identified transmission are against the rules.

Inadvertent sounds coming from your computer may not be likely to happen if you are only running a digital mode program and no other applications are running at the same time. However, to make sure there is no chance of other system sounds falsely triggering transmission, you can access the system's **Control Panel** and double-click the **Sounds and Audio Devices** icon to open the **Sounds and Audio Devices** window (**Sounds** in Win 98) and under **Schemes**, select **No Sounds**. If Window's asks if it should save the previous scheme, give your current setup a name and save it. You can then recall it at any time.

Serial Port Transmitter Keying Schematic

The circuit diagram below shows how to build a serial port to transceiver keying circuit. Your PC's serial port RTS or DTR signal line can be used to turn on a transistor that keys your transmitter. Generally RTS is used, but the *DigiPan* PSK31 program allows you to select either of the two signals.

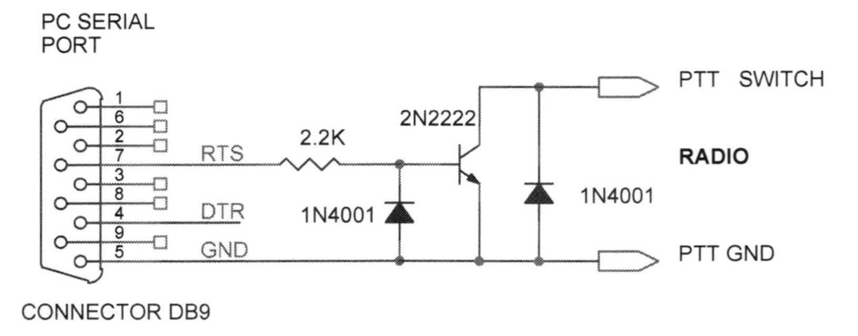

Note: On the preceding schematic, either RTS or DTR can be connected to the 2.2k resistor.

Component values in the schematic are not critical. You could substitute the general-purpose diodes and transistor for any equivalent parts you have in the shack.

You may have to do some improvising to connect the PTT control signal to your transceiver. Depending upon your radio, one of the following solutions may be the way to go:

- Connect the PTT to an auxiliary connector at the rear of your radio. If you select this approach, you may also be able to connect the sound card signals to the same connector. Consult the owner manual for your radio.
- Connect to the radio's microphone input jack. If you do this you can also connect the sound card's Speaker / Line-Out signal to the microphone input pins on the jack.
- Your radio may have an external keying jack on the back panel that can be used.

If connecting the keying circuit to your radio is inconvenient, or otherwise a problem and using VOX is unsatisfactory, you might want to consider purchasing one of the commercial interfaces discussed in Chapters 6 and 7.

Chapter 3: **Setting up DigiPan Software**

Regardless of which radio-to-PC interface you end up using, you will need a PSK31 capable program to run on your PC. In this guide we will be discussing how to use the *DigiPan* software. Other software may also be used, but for clarity of explanation, you may want to download the free *DigiPan* software.

Note: *DigiPan* requires a 266 MHz Pentium or faster PC running Windows 95 or higher operating system.

Downloading the DigiPan Software

If you have not already done so, download *DigiPan* 2.0 at:
http://www.digipan.net

To avoid potential loading problems, I usually prefer to download the zipped version: digipan20.zip. If you don't have an unzip program just click on "Download *DigiPan* 2.0 (700K)" and save the install file to some suitable location

After you have downloaded the file, locate it and click on the digipan20.exe file to install the program on your computer.

Note: If you are a Windows 2000 user, and you see a "Sound card is already in use or does not exist" error when you try and run *DigiPan*, go to **Control Panel/System/Device Manager** and look for a device called a Unimodem Half-Duplex Audio Device. Disable or remove this device and that should solve the problem.

Configuring the DigiPan Program

If the program is not already running, either click on the desktop *DigiPan* icon or find and start it by using the Windows **Start** program button.

Once the *DigiPan* program is up and running, click on the **Configure** pull-down menu and setup the following items:

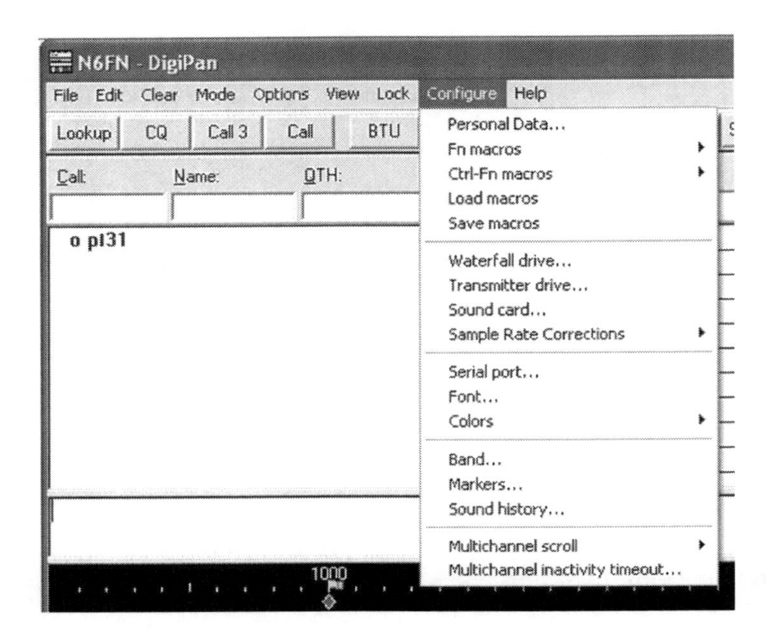

1. Configure the **Personal Data** with your **Call**, **Name** and **QTH**.

2. Verify that the **Sound card** window is set as follows:

Type:	Computer soundcard
Input:	VIA Audio (WAVE)
Output:	VIA Audio (WAVE)
Sample Rate:	11025

Note: If setting up for the *SignaLink USB* interface, the **Input** and **Output** device names will be "USB Audio Device" or similar.

3. Next select the **Serial port** that is to be used. If your PC has a serial port, select it as the **COM** port to be used. If you have a USB to serial port adapter, enter the **COM** port number for the adapter.

If you are unsure of the **COM** port number, with Windows 98, XP and Windows 7 you can access the **Device Manager** by right-clicking on the Windows **My Computer** icon and select **Properties**. Alternatively you can access the **Control Panel** and click on the **System** icon.

The following figure is the **System Properties** window for Windows XP; the similar Windows 98 screen has fewer tabs.

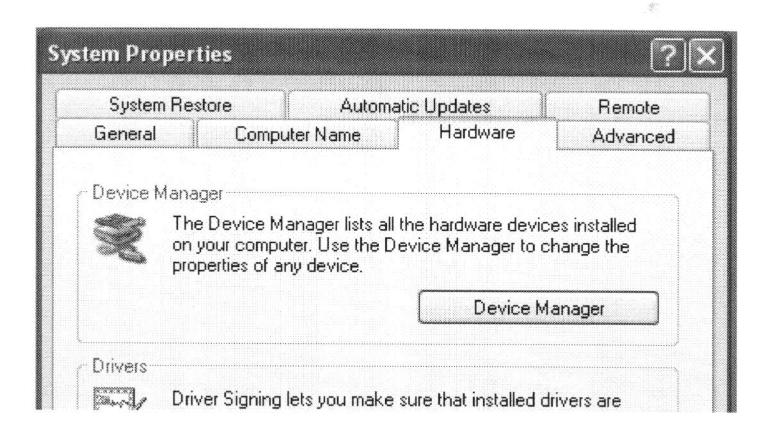

On the **System Properties** window, with Windows XP select the **Hardware** tab and click **Device Manager**. (With Windows 98 directly select **Device Manager**.) Any installed serial ports should show up under the **Ports** section, even if it is a USB to serial port adapter.

Once the **COM** port has been specified in the *DigiPan* **Configure > Serial port** window, check that both **RTS** and **DTR** are selected for PTT. That way it won't matter which one is being used in your hardware interface.

4. Click on the **Mode** pull-down menu and verify that **BPSK31** is selected.

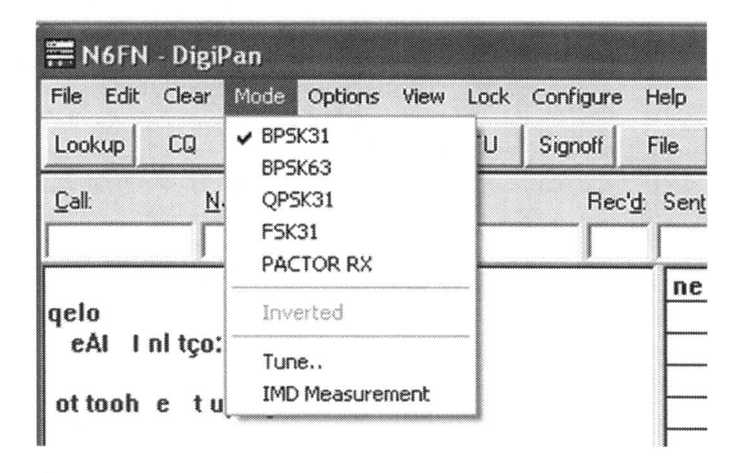

5. Next click on the **Options** pull-down menu and make sure that **AFC**, **Snap**, **Squelch** and **RX** are all selected. For now leave the other options unselected.

Setting up the DigiPan Band / Dial Scale

DigiPan is designed to provide a panoramic frequency spectrum display starting from a frequency you select and upwards up to the full width of the computer screen. If the program's window is set for a wide display, the high-end frequency could be 4 kHz or higher than the starting frequency.

For instance, let's suppose you set a start frequency of 7.070 MHz, the ending frequency could be 7.074 MHz or higher. In practice, the I/F bandwidth of your receiver limits the maximum usable bandwidth of the frequency spectrum display. Any signals beyond the receiver's I/F passband will not be detected or displayed.

DigiPan's frequency spectrum display graphically shows active signals within theI/F passband of your receiver. Amateur transceivers typically have receiver bandwidths of from about 2.5 to 3.5 kHz wide. Using DSP filtering techniques, in SSB voice mode, many modern transceivers have the capability of setting the I/F bandwidth anywhere from about 1.5 to 3.5 kHz. When operating PSK you will generally want to select as wide an I/F bandwidth as possible. In PSK operation, signal "selectivity" is provided by the PSK decode software, not by using narrow filters on the transceiver.

Since PSK31 only requires 31 Hz in bandwidth, plus some "inter-channel" guard space in-between PSK31 signals, on a transceiver with a wide bandwidth it may be possible to see 40 or more signals at a time.

Important: To be able to read the actual frequency of a signal on *DigiPan*'s spectrum display, it is essential that your transceiver be set to the same start frequency that *DigiPan* uses for the start of its frequency spectrum display.

The start frequency you select for *DigiPan* to use depends upon the operating mode of your transceiver. The standard for PSK operation is USB. If USB is selected, the spectrum start frequency to enter is the low end of the desired PSK31 band segment. This will be the same frequency you set the dial to on your transceiver.

Because of transceiver low-pass audio input frequency filtering, it is recommended that you set the spectrum start frequency 0.5 kHz lower than the first frequency of expected operation. For example, PSK31 operation on the 40-meter band generally starts at 7.070 MHz and goes up from there. Consequently we should set our transceiver and *DigiPan* at 7.069.5 MHz, which is 0.5 kHz lower.

To configure the *DigiPan* start frequencies click on the **Configure** pull-down menu and select **Band** to access the **Band properties** window. Set the band **Spectrum start** frequencies and USB mode of operation as shown below. Don't worry about the **Activate** setting at this time, as that just sets which band *DigiPan* is currently operating on.

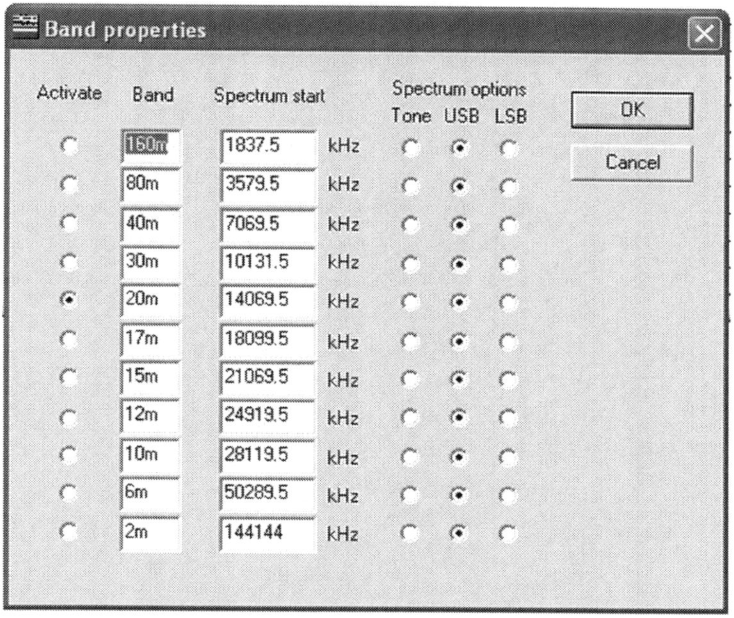

Band properties

Activate	Band	Spectrum start		Spectrum options Tone	USB	LSB
○	160m	1837.5	kHz	○	●	○
○	80m	3579.5	kHz	○	●	○
○	40m	7069.5	kHz	○	●	○
○	30m	10131.5	kHz	○	●	○
●	20m	14069.5	kHz	○	●	○
○	17m	18099.5	kHz	○	●	○
○	15m	21069.5	kHz	○	●	○
○	12m	24919.5	kHz	○	●	○
○	10m	28119.5	kHz	○	●	○
○	6m	50289.5	kHz	○	●	○
○	2m	144144	kHz	○	●	○

OK

Cancel

Alternate PSK Band Segments

Note: There is a lack of consensus among amateurs as to where PSK activity is supposed to reside on 15 meters, and on the 10, 12 and 30-meter WARC bands. On 15 meters activity may also be found at 21.080, and on 30 meters activity may be found anywhere between 10.030 and 10.040. Consequently, if you intend to operate on these bands you may want to set up the **Band properties** window differently, or perhaps just use your VFO knob to tune these areas when looking for signals.

Alternate PSK band segments summary:

- 15 meters: 21.070 or 21.080
- 30 meters: 10.030, 10.032, 10.037 or 10.040 most commonly
  Check whole segment between 10.030 to 10.040.

Setting up Your Transceiver's Receive Modes

If you have not already done so, connect the cables of your radio-to-PC interface, and using *DigiPan's* **Band** button select a suitable band. Set the following on your transceiver:

- Select USB mode of operation
- Tune to the PSK band / start frequency as shown in the **Band properties** pop-up window for the band you selected using *DigiPan's* **Band** button.
- Set your RF Gain control for maximum sensitivity. If you have lot of background noise, reduce sensitivity as required to reduce the noise while retaining low-level signal sensitivity.
- For now, turn off any Notch or Noise filtering.
  Note: Once you have things running, a modest amount of Noise filtering may improve signal reception.
- Select the widest receiver I/F bandwidth possible.
- If you have any audio filtering (bass/treble), turn it off.
- Set your radio's volume control to a normal / comfortable listening level.

Adjusting the Waterfall Signal Drive

Click on DigiPan's **Configure** pull-down menu and select **Waterfall drive**, which should display the Window's **Recording Control** window.

Note: If using Windows 95, to access the recording control window, select: **Options/Properties/Adjust** volume for recording and press **OK**.

Note: If using Vista, do not use *DigiPan*'s **Waterfall drive** setup window to access the recording level controls, as it does not work properly with Vista. Instead perform the following:

1. Plug your interface into the computer's mike jack, otherwise Vista will not recognize it.
2. Right-click the speaker icon on the Windows Task Bar near the system clock located at the lower right hand corner of your screen.
3. Select **Recording Devices**.
4. Select **Microphone** and then click the **Properties** button.
5. In the Microphone properties window, select the **Levels Tab**.
6. Adjust the **Mic** control to set your Receive Audio level.

Depending upon your operating system, use one of the above methods to locate the **Volume / Levels** control for the signal input that you are using on your computer. (Line-In, Microphone or some other input). The actual representation of the controls adjustment window will vary depending upon the operating system you are using.

Click **Select** on the control you will be using, make sure the **Balance** control is centered, and adjust the **Volume** slider to display a speckled noise pattern on the blue background of the waterfall display, with a small amount of yellow speckles showing -- similar to the pattern shown in the Waterfall Display shown below.

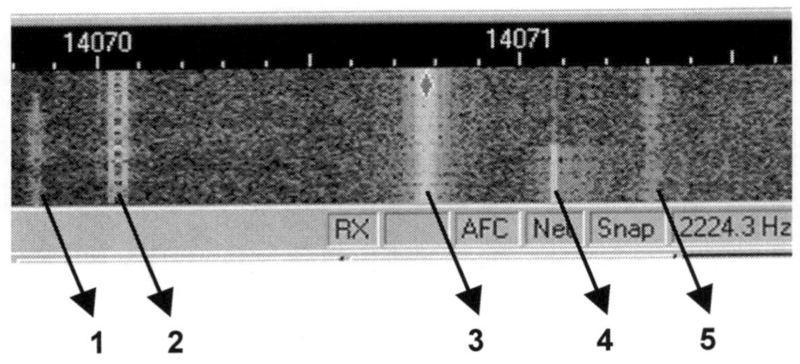

See below for waterfall display signal interpretation.

Interpreting the Waterfall Display

The *DigiPan* waterfall / frequency spectrum display contains a lot of useful information. Referring to the waterfall display above, the visible signals and frequency scale are interpreted as follows:

1. Very weak PSK station that has just quit transmitting.
2. Weak signal that still provides perfect copy.
3. Strong PSK signal which also has a diamond cursor on it.
4. Carrier, which has recently quit transmitting.
5. PSK signal that is too weak to provide reliable copy.

6. A signal's frequency can be judged by using the frequency scale at the top of the display. The minor divisions are in increments of 100 Hz.
7. The relative width of the received signal can be estimated by looking at its width relative to the frequency scale.
8. If a signal track starts showing red rather than yellow, it means that the signal is saturating the sound card input. If so and other signals appear normal, you can temporarily reduce the receiver's RF Gain to improve copy on this signal.

PSK Signal Tuning and Decoding

Tuning and decoding signals is where *DigiPan* really shines! Once your transceiver and *DigiPan* have been configured and adjusted as described in the preceding sections, it pretty much does everything automatically, even locating and decoding multiple signals at the same time.

DigiPan's software continuously sweeps through your transceiver's audio looking for decodable signals. The cumulative results of each sweep are displayed as the waterfall display grows from the top down. As sweeps accumulate, much like raster lines on a television, a picture develops which is representative of all the signals present in the transceiver's I/F passband. Strong signals are shown as bright yellow vertical traces; weak ones are shown in light blue.

One thing to remember is that you should not change your transceiver's VFO to tune to different PSK signals. So that *DigiPan's* frequency dial reads properly, leave the radio's VFO set to the same **Spectrum Start** frequency you have set in the **Band Properties** setup window and let *DigiPan* do the tuning. An exception is if you wish to tune further up the band to check for activity, especially during contests. Just be aware that *DigiPan's* frequency dial no longer is calibrated. An easy way to do this is to move up in even increments of 1 kHz, keeping a mental note of the amount of correction to apply to DigiPan's dial reading.

Now for the really great part: make sure the **AFC** and **Snap** buttons at the bottom of the screen are enabled (not grayed out), now when *DigiPan* is decoding multiple signals, just click on any desired decoded channel row at the upper right of the screen, or click on the center of the trace itself and a diamond will appear on the related signal trace. *DigiPan* then starts decoding the signal in the large decoded text window at upper left and also tunes itself for transmission on that same frequency.

Using either method, you can quickly jump from signal-to-signal at the click of a mouse.

When the **Snap** feature is enabled it will allow you to easily click to the correct tuning point of a signal trace being displayed in the waterfall. Simply hold down the left mouse button while moving the mouse cursor over the trace and the diamond marker will "snap" to the center of the trace. Perfectly tuned!

The **AFC** function serves a related tuning purpose. When enabled, any frequency drift in the trace flagged with the diamond marker will be automatically tracked by *DigiPan*, keeping the station accurately tuned.

Either or both of **AFC** and **Snap** may need to be unchecked if a nearby strong station makes it impossible to tune to a weaker station.

Multi-channel Reception

By clicking the Control Bar's **Multi** key, *DigiPan* can be operated in either single or multi-channel reception mode. Since multiple signals can be decoded at the same time, most people will probably find the multi-channel mode to be preferable.

As an aid to eliminating inactive multi-channel rows, clicking on the **Configure** key and then selecting **Multichannel inactivity timeout** from the pull-down menu brings a window where you can set a time in seconds that determines how long after no text is received passes before that channel is cleared. The default is 15 seconds.

To correlate waterfall traces to the Multi-channel decode rows, use the **View** pull-down to enable the display of A~Z **Bookmarks**.

Transmitter Settings

Set up your transceiver as follows:

- Make sure USB mode is selected.
- If your transceiver has voice transmit bandwidth controls (TBW) set them as wide as possible.
- Turn off Processor / Compression and set MIC gain normal.
- Initially set your power output to about 20% of maximum.

Setting the Transmit Frequency

Set *DigiPan* and your transceiver's VFO to one of the PSK Band start frequencies as discussed in the preceding receiver section. *DigiPan* is set to the desired band by use the Logging Bar's **Band** select pull-down menu. *DigiPan* sets the actual transmit frequency by adding a sound card created audio tone which represents the point you selected by clicking on the waterfall / frequency spectrum display.

If the **View** pull-down menu **Show frequencies** option is enabled, the actual frequency offset added by *DigiPan* is easily seen in the **RX** window at the bottom of the screen.

As you will recall from the prior section, because of low-pass audio frequency filtering within the transceiver, the PSK band frequencies were set 500 Hz lower than where we expected to receive signals. For the same reason, when choosing a place to transmit avoid using the first 500 Hz of displayed spectrum.

In addition, you should limit your transmissions to several hundred Hz below your maximum USB transmit bandwidth. If you know what the transmitted SSB audio filtering high-cutoff frequency for your transceiver is, stay 300 to 500 Hz below that. If you don't know, if should be fairly safe to limit your transmission to 2100 Hz above the set PSK Band start frequency.

If you have a recent vintage transceiver, it may have either selectable / multiple transmit bandwidth filters, or the bandwidth may be DSP processor adjustable. Use the widest settings available.

Important: Keep the transmit frequency offset, as seen in the **RX** window at the bottom of the screen, within the minimum and maximum frequencies discussed above. In general this will be no less than 400 or 500 Hz on the low side and depending upon your transceiver's transmit bandwidth, from 2100 to 2400 Hz on the high side.

Note: The width of the blue and yellow speckled area of the waterfall display as a good approximation of your receiver's bandwidth. The area where the waterfall display is black or lacks very much speckling is outside the effective bandwidth of your receiver. In addition to the above limitations on your transmit bandwidth, you will want to keep transmissions within the speckled area – your effective receive signal area.

Adjusting the Transmission Drive Level

There are two primary issues to be concerned about when setting your transceiver's output power and drive level from the sound card:

1. Since the duty cycle of PSK is greater than SSB, to protect your power amplifier you probably don't want to run your rig at maximum power. A power output setting of about 25% of maximum should satisfy this requirement. Don't be concerned about not running full power, PSK is very efficient and you can work lots of DX with ten watts. In fact, running high power can cause interference to near-by stations

2. From the standpoint of transmit signal quality; you need to make sure that there is no ALC (automatic level control) action while transmitting. Any ALC activity will likely cause distortion of your signal (resulting in poor IMD).

Using the *DigiPan* **Configure** pull-down menu, select **Transmitter drive**, which brings up the Window's **Volume Control** window as shown on the following page. As an alternative, on most systems double-clicking on the small speaker at the lower right of the PC's screen should also bring up the **Volume Control** window. (See following page for instructions using Vista.)

Initially set the master **Volume Control** and **Wave** sliders to mid-range. Also to prevent any inadvertent signal pickup from other signal sources on your PC, **Mute** any other devices shown on the screen. Refer to the following screen shot.

The combination (sum) of the **Volume Control** and **Wave** controls sets the level being sent to your transceiver.

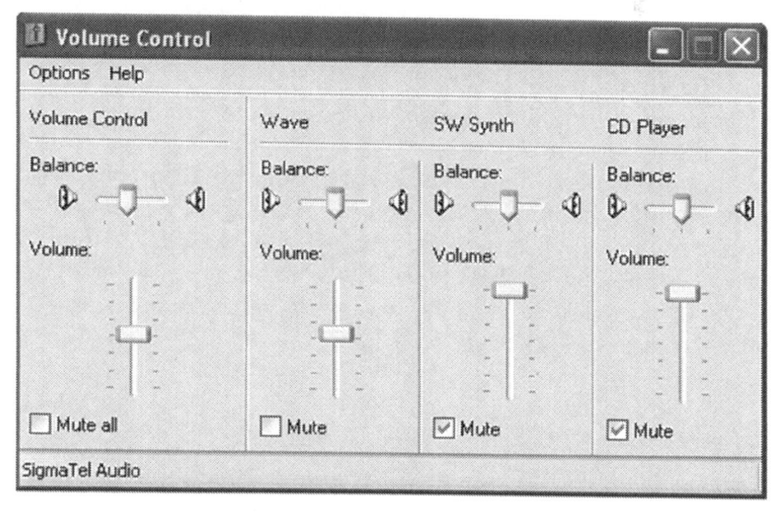

Note: If using Vista, do not use *DigiPan*'s **Transmitter drive** setup window to access the recording level controls, as it does not work properly with Vista. Instead perform the following:

1. Plug your interface into the computer's mike jack, otherwise Vista will not recognize it.
2. Right-click the speaker icon on the Windows Task Bar near the system clock located at the lower right hand corner of your screen.
3. Select **Open Volume Mixer**.
4. If you have a USB audio device like the SignaLink USB, click **Device** and select **Speakers (USB Audio Codec)**, otherwise select your computer's sound card.
5. When DigiPan is transmitting, another volume control will show up as either **Tune TX** or **Callsign-DigiPan**.
6. Use a combination (sum) of these two controls (3&4) to set the Transmit Drive Audio level, as described above.

If your radio-to-PC interface has an adjustable transmit level control, start off by setting it as recommended by the manufacturer. Both the BuxComm *Rascal GLX* and TigerTronics *SignaLink USB* units discussed in later chapters have such a setting.

At this point we are ready to transmit. If available, use a dummy load while making the following adjustment.

To start transmission click on the **T/R** button in the Control Bar at the top of the screen, both the **TX** window at the bottom of the screen and your transceiver should indicate that you are transmitting.

Watching your **ALC** meter, reduce the **Volume Control** level until you see no **ALC** activity on the meter. Slide the control up-and-down to locate the point where you are just below where **ALC** activity starts to show on the meter. This will be the maximum drive level setting that can be used that is consistent with low signal-distortion and minimal transmitted sidebands.

Inadvertent Sound Card Level Changes

Once you have set the sound card volume levels for transmit and receive, don't be surprised if they later get changed. As you use your computer to run other programs, some programs may automatically change the settings or perhaps you may change them yourself to listen to music, view video clips or run other digital modes. So be prepared, the sound card input and output level settings you so carefully set are likely to be altered.

By first making a note of your PSK volume slider settings, after using your computer for other purposes, it is then just a matter of accessing the playback and recording controls and resetting them back to where they need to be whenever you want to run PSK again.

Note: If using Vista, you should not have trouble with the drive level setting being changed by other programs. Vista generates different volume controls for each software application that is running. Consequently, you should not need the RoMac *Sound Card Manager Program* described below.

There is a hardware way of avoiding this problem. By using an interface that has its own built-in sound-card chip, and RX / TX and volume controls, like the TigerTronics *SignaLink USB* interface, it's impossible for your PSK volume settings to be inadvertently changed. If you are using the *SignaLink USB*, you will not need the *RoMac Sound Card Manager* program described below.

KF8OY's RoMac Sound Card Manager Program

If you are not using an interface with its own receive and transmit volume controls, like the TigerTronics *SignaLink USB,* KF8OY's *Sound Card Manager* program is a rather elegant solution to the volume setting problem, especially if you use your PC to operate several different digital modes. It's pretty much a given that different digital modes will require their own sound card input and output level settings.

Roger's program works on all Windows machines, including Vista, and is designed to remember the sound card settings and reload them automatically when you restart your digital modes, or any other programs. When you close your application, the *Sound Card Manager* program restores the original settings. Very neat! The program is available as freeware from Roger's web page:
http://www.romacsoftware.com/SoundManagement.htm

If you use multiple programs that can affect the sound card level settings, I would recommend using *Roger's Sound Card Manager*. In addition to keeping track of the sound card settings for multiple applications, it makes finding and setting the Window's Record and Play levels a breeze.

RoMac Sound Card Manager Program Setup

By way of setup, get the program started and decide if you want the program to be run from the Windows system tray at the lower right, or if you would prefer to have it start when you double click on its desktop icon. If you wish to have it start in the system tray, select **Start Minimized in Tray**.

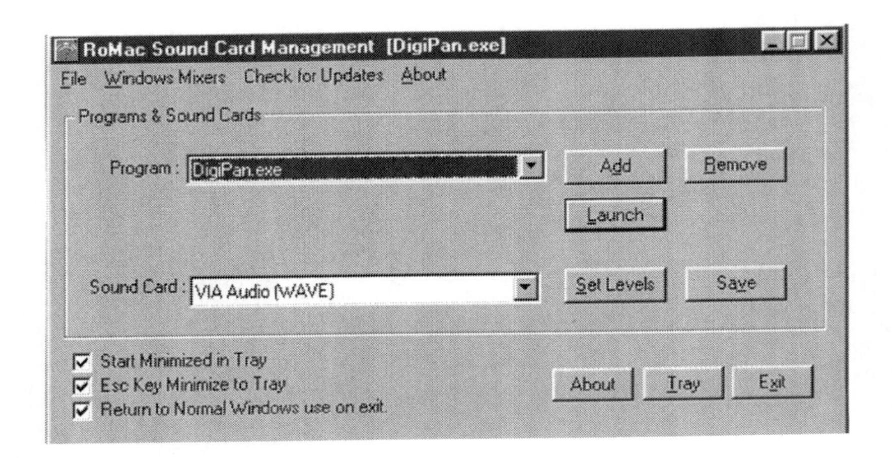

After that, use the **Add** button to add all your "sound" applications to the program list and specify which sound card you want used with that application.

Whenever you want to run an application that uses the sound card, first start the *Sound Card Manager* program. Then click the **Launch** button to start the desired application. The first time the application is launched, set the desired sound card levels by clicking *Set Levels*. After setting the levels, **Close** the adjustment window and **Save** the settings. After that whenever you again **Launch** that application, the sound card levels you saved will be retrieved.

If the **Return to Normal Windows** on exit option is enabled, when you close the *Sound Card Manager*, the original Windows levels will be restored.

Chapter 4: **Operating DigiPan**

Once you have decided on an interface, installed it and everything is setup as described in the preceding chapter, you are ready to start making contacts. During your first contacts you can verify transmit signal quality by asking for signal reports.

Clearing Excess Display Clutter

By now you probably have noticed that the multi-channel decode text and waterfall screen start getting a bit messy after a while, with multiple signals being bookmarked and decoded, some of them no longer transmitting. The solution is to clear all the displays by clicking on **Clear** in the Control Bar.

If you don't want to clear everything at once, you can selectively clear items by clicking on the top row **Clear** pull-down menu and selecting one of the following:

- **Clear RX window** (the larger decode window on the left)
- **Clear multichannel window**
- **Clear TX window** (the text window just above the waterfall)

It is also possible to selectively clear individual rows of the multi-channel display by moving the cursor over the row to be cleared, right-clicking and then selecting **Clear Channel x**.

Looking for stations calling CQ

Instead of searching through decoded text in the multi-channel decode area; you can look for a red background, which indicates stations calling CQ. Look for a signal of reasonably good strength, click on its associated multi-channel row and you are tuned to the frequency and ready to transmit.

If a multi-channel station calling CQ does not show a red background, refer to "Customizing the Multi-channel Coloring Effect" on page 35.

It gets better. By double-click on the calling station's call sign in the main decode window; it is copied to the **Call** field, ready to be used when you make your call to the station. Likewise, if you double-click on the name of the person calling CQ it is copied to the **Name** field.

Responding to a Station's CQ and Sending Text

Once you have the station's call recorded in the **Call** field, clicking on **Call 3** sends out a call to the station in the following format:
CALL CALL CALL de MYCALL MYCALL MYCALL

The text of your **Call 3** message immediately appears in the transmit window and is progressively underlined as it is being transmitted. If transmission does not automatically stop after all the text has been sent, press the **T/R** button to stop transmission.

If the **Call 3** button did not automatically terminate transmission after the call had been sent, right-click the **Call 3** button and edit the macro seen in the window by adding the <RXANDCLEAR> macro command after the last <MYCALL> in the sequence.

When a station comes back to your call and you are ready to answer, click on the **Call** button, which will both enable transmission and send your call in the following format: CALL de MYCALL. This time however, transmission does not terminate and any text typed in the transmit window will be sent. As text you type is transmitted, it is also copied to the large receive window. As a result, the receive window contains a complete record of both sides of the conversation.

If desired, you can type text in the transmit window while still receiving text from the other station. This text will not be sent until you click on the **T/R** button, which will then start transmission.

Text in the transmit window may be edited via the normal Windows editing functions before it is sent out:

- Backspace key to delete characters to the left of the cursor
- Delete key to delete characters to the right of the cursor
- Left and Right arrows to move the cursor within a line
- Up and Down arrows to move between different lines

After you have finished replying to him by typing in the transmit window, press the **BTU** button to send your call and terminate transmission. Your closing call is sent in the following format:
btu CALL de MYCALL K and then transmission terminates.

At this point you can continue the QSO as usual. If this is one of your first contacts, you might want to ask for a signal quality report. In addition to receiving a signal strength report, also ask if your transmission bandwidth looks good. (If it is too wide, or has harmonic bands showing, it's probably caused by excessive transmission drive level. If so reduce drive as discussed earlier.) He may also give you an IMD report, which is discussed in the following section. If he gives you a RSQ report (as opposed to RST), refer to Chapter 5 "The RSQ Reporting System" if you are not familiar with the digital modes RSQ reporting system.

On your final transmission, when signing with the other station, you can press the **Signoff** button and the following will be sent, after which your transmission will terminate.
73 CALL de MYCALL SK

Note: Any of the message text and command strings in the macros used in the above-described buttons can be edited by right clicking on the associated button. A window opens, displaying the body of the macro on the left, and on the right additional commands that could be inserted into the macro. See the **Edit user macro** screen shot on page 36.

If you wish to edit the macros defining the operation of the various buttons in the Control Bar, refer to page 36 and the **Macro Programming** section of the *DigiPan* **Help** screens.

If the transmit window is too small (not tall or wide enough) use your mouse to click on the bottom edge or sides of the window and drag it as desired to enlarge the window.

You may occasionally want to clear old text out of the transmit window, if so, press the **Clear** button and select **Clear TX window** from the pull-down menu.

If you wish to manually terminate transmission, press the **T/R** button and after any remaining text in the transmit window has been sent, transmission will cease. On the other hand, if you wish to terminate transmission immediately press the **ESC** key on your computer, or alternatively click on the **RX:** cell in the Status Bar at the bottom of the screen. Transmission stops immediately.

IMD Measurement

IMD (Inter-Modulation Distortion) is an important PSK signal quality measurement, which is commonly used in PSK communication signal reports. It helps operators ensure that their PSK31 setup is transmitting properly. An IMD measurement capability is a commonly found function in most PSK programs. In *DigiPan* you will see the measurement in the **IMD** cell at the bottom of the screen.

IMD is measured when the system is "idling" in transmit mode, with no characters being sent. While idling, the system is transmitting a 31 Hz two-tone signal. A receiving station's software makes a measurement of the signal's 62 Hz harmonic, comparing it to the strength of the main signal. The difference in signal strength between the main signal and the 62 Hz harmonic is then expressed as a negative dB number. The more negative the better.

IMD measurements in the range of –25 to –30 dB are considered very good. Readings of –20 dB or greater (a number smaller than 20) are considered poor and are usually caused by excessive transmitter drive level. If you are getting less than good readings, try reducing the drive level from the sound card.

Getting and Giving IMD Reports

Once another station is ready to give you an IMD report, let him know when you will place your station in idle mode. Let it idle for ten or fifteen seconds, allowing him enough time to make the measurement.

If someone asks you to make an IMD measurement, wait for his signal to be in "idle" for several seconds and then make the reading using the Status Bar's **IMD:** cell. Make a note of several readings, reporting an average. If necessary, click the cell to force a reading.

In addition to excessive transmitter drive level; the following variables can also contribute to poor IMD readings:

- Weak signals, for best results the signal should be reasonably strong.
- Excessively strong signals (with red tracks)
- Other signals in close proximity to the signal of interest
- Noise or other signals being mixed with the transmitted PSK tone. These may be caused by signal pickup in your cables, a noisy sound card or other problems.
- Lack of good grounds between your equipment.
- If using an un-isolated interface cable arrangement as described in Chapter 2, ground loops between the PC and transceiver may be causing problems. If so try using a different interface that provides isolation. (In this book, both the *Rascal GLX* and *SignaLink USB* interfaces are isolated.)

Enabling and Setting DigiPan's Signal Decoder Squelch

When the Status Bar's **Sq** button is enabled, only signals above a set threshold will be decoded. On weak signals this helps prevent random noise from being detected as "garbage" characters. This is analogous to using squelch on an FM receiver. Clicking the Control Bar's **Squelch** button brings up the **Squelch** level adjustment window.

Be careful not to set the squelch level threshold too high, as it may prevent readable weak signals from being decoded. Whenever the signal is below the set threshold, the Status Bar's **Sq** button will turn red. For weak signal reception, it may be best to turn off the squelch.

Customizing the Multi-channel Coloring Effect

As noted earlier, upon decoding a "CQ" signal, *DigiPan* can be configured to turn the multi-channel row background red or some other color. Background colors can be configured for up to three sets of text that you define. This can be put to advantage to flag calls being made as part of a contest, a friend's call sign or anything else that you might be looking for.

To review or change the current settings, click the **Configure** button and select **Colors** from the pull-down menu. This will bring up another menu from which you should select the **Multichannel coloring effect** entry, which displays the search-string text and background color selection window. You can modify the text **Strings** to search for signals containing any specific text sequence you desire.

Sending Station Info with the Brag Button

The Brag button can be used to format and send any text you like. This could be your signal report, a statement of your name, station equipment being used, QTH and so forth. You could also redefine both the button's name and the contents of the macro to be anything you like.

Assuming you wish to use the **Brag** button to give the other station information regarding your station, right clicking on the button will open the **Edit user macro** window where you can customize the button's name and its contents as desired.

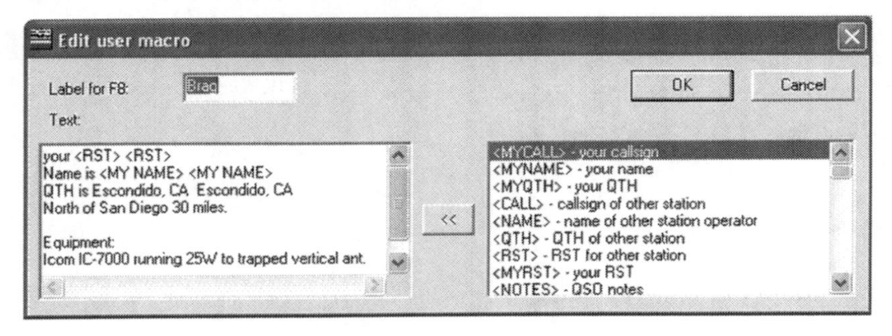

Try and minimize the use of UPPER CASE characters and special symbols, since they have more "bits" in their codes, they take longer to transmit. Once you have configured the **Brag** button macro as desired, it's a simple matter of clicking the button any time you wish to send your station information.

Creating Custom Button Functions

In addition to the button functions predefined on the primary Control Bar, there is a secondary set of buttons that allow you to create custom functions of your own. Click the " ^ " button at the far right of the Control Bar to toggle between the primary and secondary Control Bars. Alternatively, while holding the **CTRL** key down on your PC the secondary Control Bar is enabled, which you may find more convenient to use.

Right clicking on any of the Control Bar buttons will open the **Edit user macro** window where you can customize the button's name and its contents as desired.

One neat way of using this during contests or special events would be to create one button to send out a contest or event specific CQ, and then also create a second button which would send the contest's standard exchange or a "brag tape" of the purpose of the special event.

By using the macro commands in the right side of the **Edit user macro window**, you can also enter QSO specifics, such as the station's call, name, report, QTH etc. In addition there is a set of control functions that can be useful, depending upon your creativity, for starting and ending transmission, clearing one or more of *DigiPan's* windows to get ready for the next QSO and others. There is even a command, which allows you to transmit a file on your PC, **<FILE>**. When this macro runs (is executed) a window pops-up where you can select the file to be transmitted. If instead you wish to imbed the file name into the macro itself, use the **<FILE:*FILENAME*>** form of the command, substituting the actual filename in the ***FILENAME*** field.

Calling CQ

Calling CQ is extremely easy. Just click on the Control Bar's **CQ** button, transmission is automatically enabled and your CQ is sent out in the following format, after which transmission will terminate so you can "listen" for a response.
CQ CQ CQ de MYCALL MYCALL MYCALL

Using the Built-in Logger

While *DigiPan* has some great features and is fun to operate, its built-in logging capability is rather limited. Although, it is better than a using a manually generated paper log because it can easily capture the station's call, QTH, signal reports, frequency of operation and the date as part of making the QSO. While casual PSK31 operators may find the logging capability adequate, contester's and award chasers will probably want to do their logging on a different logging program or use one of the logging programs that has built-in PSK operation capabilities.

Once you have found a signal of interest, or a station has come back to your call, double-clicking on his call in the main receive window will copy it to Logging Bar's **Call** data entry field. Likewise, double-clicking on the station's name will cause it to be entered in the **Name** field. Additionally, any of the Logging Bar's data-fields can also accept manual entry by clicking on the field and typing in it.

If you wish to see if you have previously contacted this station before, press the search button, the eyeglass icon, and a pop-up window is displayed showing the current station's call. Pressing **Search** will bring up a window showing any previous log entries with this call.

The QTH can be entered by using the mouse to highlight the transmitting station's QTH information, right clicking the mouse and selecting **Copy** from the pop-up window. Then move the mouse cursor to the QTH entry field and right-click again, selecting **Paste** to copy it into the data-field. Alternatively, you can enter the station's QTH word-by-word by holding down the PC's **Shift** key and double-clicking on successive words of the QTH. If the station has sent his QTH multiple times, this latter alternative will be preferable when there are decoding errors in some of the displayed QTH text.

The RST or RSQ report you have received can easily be entered into the **Rec'd** data entry field by double-clicking on the appropriate 3-digit field in the receive window. Refer to page 43 for information on the digital modes RSQ signal quality reporting system

If you enter the other station's RST or RSQ report into the **Sent** cell's data-field before sending him his report via the **Brag** button or any other button that uses the macro <RST> command, it will automatically be sent when the button is pressed.

You can freely type any notes and comments into the **Notes** data-field.

Remaining log fields (those not visible in the logging bar) record the time, frequency and mode and are automatically filled in by *DigiPan* when the log entry is saved.

To save your log entry, press the save button, which is the floppy disk icon. The little asterisk " * " in front of this button indicates that the log entry has not yet been saved. Once it has been saved, the little asterisk will disappear.

When you are ready to start another QSO, double-clicking on the next call sign, or pressing the white paper sheet icon clears the current Logging Bar entries.

Examining, Editing or Deleting Log Entries

If you wish to examine or edit the entries in your log, press the search button, the eyeglass icon, and when the pop-up window is displayed, select **Whole log** and the log is displayed as shown below.

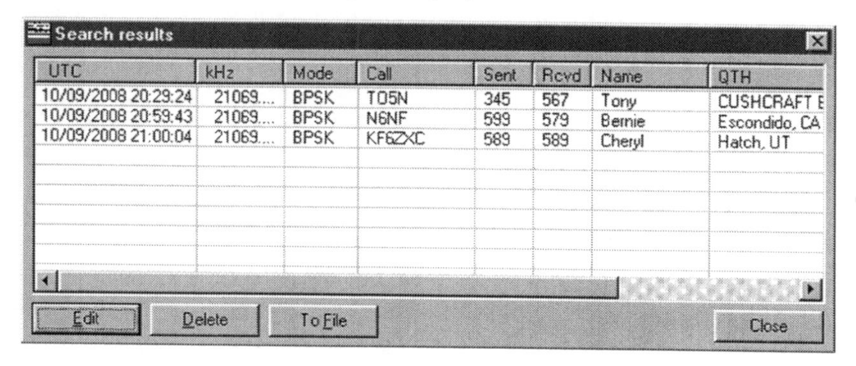

To edit or delete entries in the log, click on either the **Edit** or **Delete** buttons.

Printing and Saving Log Entries to a File

If you wish to print the log, or a portion of the log, first display the whole log as described above, then use the mouse to highlight the desired entries. To select a range of entries, place the mouse cursor at a log entry at one end of the range and click, then move the mouse to the other end of the desired range, hold down the PC's **Shift** key and click the entry – the whole range will then be highlighted in blue.

You can also select multiple non-sequential multiple log entries by clicking on the individual entries while holding down the **Ctrl** key on your PC.

Pressing **ToFile** brings up a pop-up window where you can name the file. Pressing **OK** saves the file in the *DigiPan* folder.

If you wish to print the file, or perhaps prepare it for export to another logging program, access your saved log file in the same folder that that contains the *DigiPan* program. Double click on the desired file and your default text editor will open the file for printing or editing.

Displaying Local or UTC Time

The current time, derived from your computer, is displayed in the Status Bar at the lower-right corner of the *DigiPan* window. By accessing the **View** pull-down menu at the top of the screen you can set the time to be displayed in either **UTC** or local time formats. If the time on your computer is already set in UTC instead of local time, the Windows **Time Zone** setting needs to be set to GMT and adjustment for **daylight savings changes** disabled. To access your Windows clock setup screen, right click on the time clock showing at the bottom right of your PC's screen.

Call Sign Lookup

If your computer has an active connection to the Internet, the current call sign in the log entry's **Call** data-field can be looked up at either QRZ.COM or at HAMCALL.NET by pressing the control bar's **Lookup** key. First make sure that under the **Options** pull-down menu that **Call lookup** is configured for **QRZ.COM** or **HAMCALL.NET** as desired..

Other DigiPan Modes of Operation

In addition to PSK31, *DigiPan* has other modes of PSK operation that you may want to try. However, to avoid interfering with other stations operating PSK31, only use these other modes in their appropriate band segments.

For more detail on these modes refer to the **PSK31 Code Variants** section in Chapter 1.

If you wish to try some of these modes of operation, if for nothing else to see if you can receive any compatible signals, use the **Mode** pull-down menu to make your selection.

- **BPSK31** The normal mode for PSK31 operation at 31.25 baud.
- **BPSK63** Higher speed PSK operation at 62.5 baud.
- **QPSK** Quadrature Phase Shift Keying, with error correction.
- **PACTOR** Only supports pacTOR-1 reception, no transmission.
- **FSK31** An experimental narrow frequency FSK mode.

Of these, perhaps BPSK63 is the most used. It is often preferred by contesters and others who appreciate the faster data rates which speed up their operations.

Because PSK63 is twice as wide as PSK31, to help reduce band congestion it is recommended that you operate on even 100 Hz frequencies. The suggested area for PSK63 operation on 20 meters is just above the PSK31 activity, starting at 14072.5 and extending to the beginning of the RTTY activity area at 14080.

To do this, just set your transceiver dial frequency to 14072.0 and operate at *DigiPan* spectrum display frequencies in even increments of 100 Hz, starting at 500, 600, 700, ...etc. Of course, if you are answering someone else's call, just point and click on any PSK63 signal you see on the waterfall.

Suggested frequencies for PSK63 operation on the low bands start 2.5 kHz above PSK31 activity, and extend upward for 2.5 kHz in even 100 Hz increments, starting at 500 Hz. In general, based on current PSK31 band segment usage, PSK63 activity should start at:

- 3582.5 on 80m,
- 7037.5 or 7072.5 on 40m,
- 10142.5 on 30m,
- 14072.5 on 20m,
- 18102.5 on 17m,
- 21072.5 or 21082.5 on 15m,
- 28122.5 on 10 m.

Set your transceiver to these start frequencies and use *DigiPan's* spectrum display to select operating frequencies in 100 Hz increments starting at 500 Hz.

For the latest information on operating PSK63 refer to the following URL, which is maintained by Skip Teller, KH6TY, one of the principal designers of *DigiPan*.
http://www.qsl.net/kh6ty/psk63/quikpsk.html

You will notice, that PACTOR is supported as a receive only mode. The designers of DigiPan did this so that if PSK operators suspect they are being interfered with by PACTOR stations "taking over" a PSK frequency already in use, they can identify the source of the offending signal. I personally have not experienced this, but apparently this has been a problem in the past.

Chapter 5: **The RSQ Reporting System**

Several years ago the IARU endorsed the RSQ reporting method for IARU Regions 1 and 3, which includes the Americas, Europe and just about all the rest of the world except for the Asia Pacific Region. The move to transition from the traditional RST system, which was developed for CW, to the new more meaningful digital-modes RSQ system has been going on for quite some time now. Many amateurs around the world and in the USA are using the new system

Using the new RSQ Reporting System

The RSQ signal reporting system is structured much like the RST system we are all familiar with. RSQ stands for Readability, Strength and Quality and is issued in a 3-digit format, exactly like RST reports.

RSQ Readability: For digital-mode operation, Readability ratings of from 1 to 5 indicate the percent of good text received and is quite descriptive of a signal's effectiveness.

RSQ Strength: Because digital mode communication often has multiple active signals present in the receiver's pass band, S-meter readings, being cumulative readings, are not representative of any individual signal. Instead of relying on S-meter readings, the RSQ system uses a visual interpretation of the signal trace as seen in the waterfall display to measure the signal's strength relative to the background noise level.

RSQ Quality: Since tone, hum and clicks quality assessments are relatively meaningless for digital modes of communication, the RSQ system relies on a quality assessment of the signal's waterfall trace, looking for any unwanted modulation, splatter or harmonics present on the signal.

RSQ Ratings Defined

Examining the following table, you will see a structure similar to the RST system used for CW signal reports, but with the emphasis on how well text is received and using the waterfall display to make signal quality assessments.

Readability (in % of text)

- **5** 95%+ Perfectly readable text
- **4** 80% Practically no difficulty, occasional missed characters
- **3** 40% Considerable difficulty, many missed characters
- **2** 20% Occasional words distinguishable
- **1** 0% Unreadable

Strength

- **9** Very strong trace (notice there are only 5 of these)
- **7** Strong trace
- **5** Moderate trace
- **3** Weak trace
- **1** Barely perceptible trace

Quality

- **9** Clean signal, no visible sidebar pairs
- **7** One barely visible sidebar pair
- **5** One easily visible pair
- **3** Multiple visible pairs
- **1** Splatter over much of the spectrum

Note:

The following URL is a link to an article written by Graeme Harris, VK3BGH, one of the developers of the RSQ reporting system. The article explains the rational for the system and has examples of good and poor quality PSK31 waterfall signals.

www.psb-info.net/IARU/RSQ-improved-signal-reporting-for-PSK.pdf

Chapter 6: **Rascal GLX Radio Interface**

From a functional standpoint, the *Rascal GLX* interface made by BUXcomm is essentially an improved version of the homebrew interface described in Chapter 2. While our home-brew interface had the Line-in and Line-out signals wired directly to the radio, the *Rascal GLX* includes isolation transformers to provide protection from potential ground loops. That is a definite plus.

Several other enhancements are included on the *Rascal GLX* module:

- A LED, which indicates when the unit is in transmit mode, helpful if you need to troubleshoot TX keying problems.
- A potentiometer providing additional transmit drive control.
- An additional RTTY mode keying transistor for use with radios that come equipped with an auxiliary RTTY / FSK keying input.
- Comes equipped with a ferrite to reduce noise pickup on the sound card cables.

For comparison to our home-brew interface of Chapter 2, below is a simplified representation of the *Rascal GLX* schematic showing the added Line-in / Line-out isolation transformers and the variable gain potentiometer for additional transmit drive control.

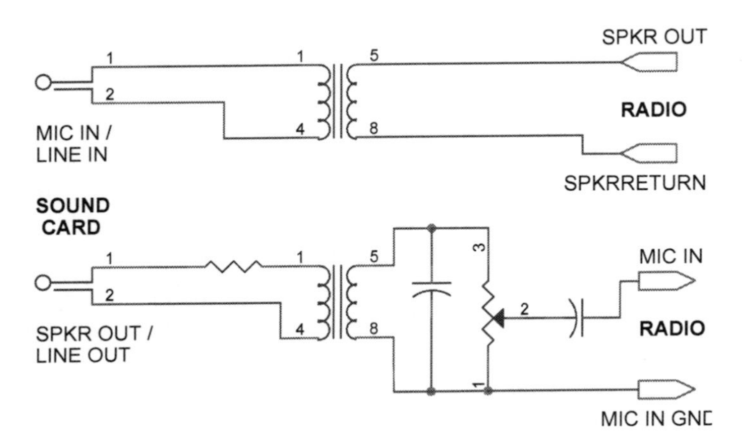

Like the home-brew interface of Chapter 2, the *Rascal GLX* requires a serial port for keying the transmitter. If your PC only has a USB port you will need a USB to Serial adapter. These are available from a variety of sources, including BUXcomm.

Again, for comparison to the home-brew keying interface of Chapter 2, the schematic below shows two diodes being used so that either or both of the RTS and DTR signals can be used to key the transmitter. You can also see the transmit indicator LED, which is in series with the common side of the two input diodes.

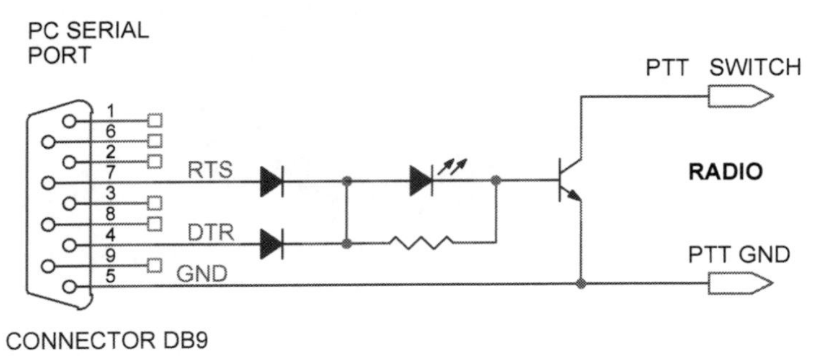

Advantages of the Rascal GLX Interface

In addition to the features mentioned above, one of the big advantages that the *Rascal GLX* has over the home-brew interface, is that BUXcomm provides a variety of interface cables for plug-and-play compatibility with a wide variety of amateur transceivers. This saves you a lot of research and experimentation, not to mention having to find the needed connectors. When you purchase your unit, it comes with an interface cable that you select. Having multiple cables available is especially nice if you have more than one radio you want to use the interface with. Their webpage allows you to look up which cable you need by the type of radio you have.
http://www.buxcomm.com/

There are a number of other companies providing similar interfaces and self-assembled kits, which may work just as well. The *Rascal GLX* was picked as being representative of this class of interface and alternative products will essentially be setup and operated in the same manner. The very popular *RigBlaster's* from Western Mountain Radio, and a variety of interfaces from MFJ, while offering additional features, are essentially similar when it comes to the sound card to transceiver interface methodology.

Rascal GLX Installation and Setup

Setting up *DigiPan* for use with the *Rascal GLX* is exactly the same as for the home-brew interface of Chapter 2. The only difference is that the *GLX* has a potentiometer for additional control in setting the transmit drive level. To start with, just center this potentiometer to mid-range. After centering the control, proceed as described in Chapter 3, "Setting up *DigiPan* Software".

During setup, if you have insufficient adjustment range when setting the transmit drive level using the Windows **Volume Control** window settings as described on page 26, use the *Rascal GLX's* level control to either decrease or increase the signal as required.

Chapter 7: **SignaLink USB Radio Interface**

TigerTronix has been providing sound card interfaces for a number of years and the *SignaLink USB* is their flagship offering. I consider it to be a third-generation design. While first-generation interfaces were essentially not much more than directly connected cables like the homebrew interface of Chapter 2, and second-generation products incorporated other features and a variety of improvements, but were still "tied" to the PC's sound card and the various problems that can bring. To my way of thinking, third-generation products would be those that include an Audio Codec or other device, the equivalent of a "sound-card chip," in the interface hardware, divorcing the interface from the PC's sound card.

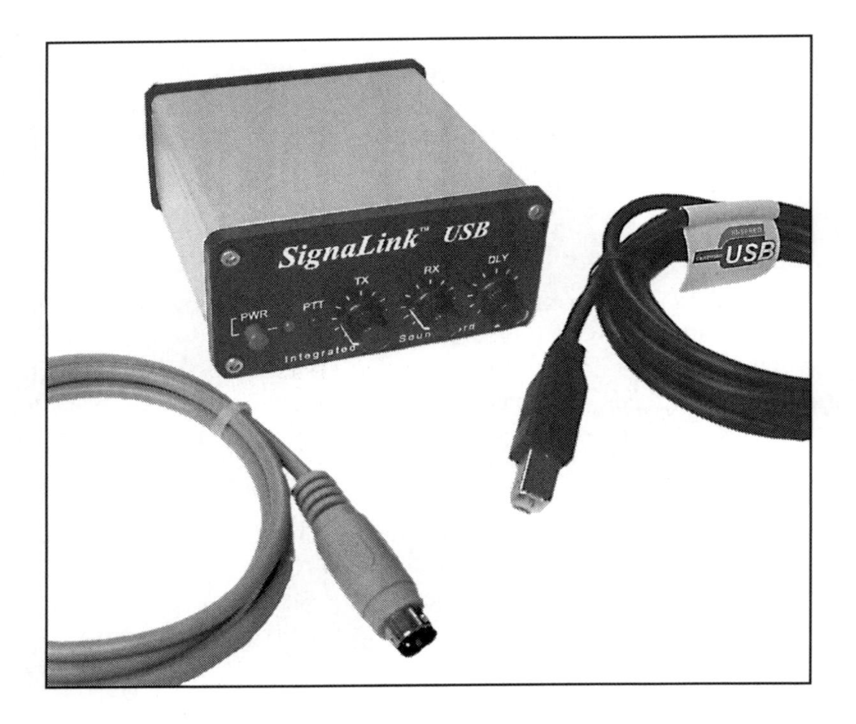

Advantages of the SignaLink USB Interface

The design of the *SignaLink USB* is an attempt at resolving a number of problems that can come up when trying to interface a PC's sound card to an amateur radio transceiver. While a number of these problems are also addressed by the various second-generation products that are available, the TigerTronics *SignaLink USB* goes a step further. The following list highlights some of the problems that the *SignaLink* is designed to eliminate or reduce:

- Freedom from PC generated sounds causing interference.
- Independence from PC applications altering the RX and TX sound level adjustments.
- Isolation from ground loops causing transmit signal distortion and subsequent difficult to diagnose poor IMD readings.
- With its own low-noise audio codec, independence from the varied quality of PC sound-card sensitivity, noise and fidelity.
- Modern PC's no longer include serial ports.
- PC serial port conflict and configuration issues.
- Internal VOX eliminates PC generated keying issues.
- Powered via the USB cable, eliminating a "wall wart".
- Ease of adjusting TX and RX drive controls on-the-fly without having to open a PC window volume control.
- Support for multiple transceiver brands and models via a selection of pre-made radio interface cables and jumper configurable signals within the *SignaLink* box. A jumper header allows a single cable to be used with multiple radios.
- Simplified interconnect cabling, using a standard USB cable and a pre-made transceiver interface cable.

The SignaLink USB is a good alternative if you don't have a serial port available and would like the simplicity of installation it provides. As of the time this guide was written the list price for most models of the *SignaLink USB* on the TigerTronics web site was $109.95, including your choice of pre-made radio interface cables. In case you need additional radio interface cables, most of them were shown as $19.95 each. More information is available on their web site: **http://www.tigertronics.com/**

SignaLink USB Functional Block Description

From a functional standpoint, the *SignaLink USB* is interconnected between the PC and your transceiver as shown below. The only connection to the PC is via the USB cable, which also supplies power to the *SignaLink's* circuitry. Connection to your transceiver's auxiliary data jack is via a single pre-made cable.

.

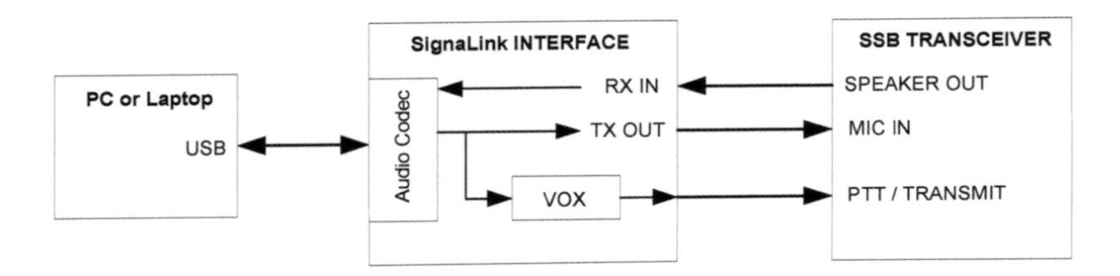

SignaLink PC to Transceiver Interconnect Diagram

Various pre-made cables are available for all radios that have a Data or Accessory Port that uses a 5-pin DIN, 8-pin DIN, 13-pin DIN, or 6-pin mini-DIN connector. Auxiliary connector pin-out variations within a connector type are handled via a configurable jumper block internal to the *SignaLink.* That way the same cable can be used with multiple transceivers, even though their connector pin-out configurations may be different.

If your transceiver does not have an auxiliary rear-panel jack available, cables are available for all radios that use a 4-pin round, 8-pin round, RJ-11, and RJ-45 microphone connectors. If using one of these, an additional audio cable (included), is used for connecting to the transceiver's speaker output.

In the preceding diagram, notice that the transceiver's PTT / Transmit keying is done via a VOX circuit internal to the *SignaLink.* The VOX activates transmission anytime the TX-out signal is of sufficient amplitude, thus eliminating a cable to the PC's serial port. Note: this feature is not using your transceiver's VOX, which should be turned off.

Receive signal detection (waterfall drive) in the *SingaLink USB* is not dependent upon any level control settings on your PC. The level is controlled via the RX knob on the front-panel. On the transmit side, after presetting your PC's volume control to a suitable gain setting, all future drive level adjustments can be made using the front-panels TX knob.

The front-panel also comes equipped with a transmit indicator LED and a DLY knob which sets the VOX hold time, similar to the VOX delay adjustment on an amateur transceiver.

SignaLink USB Installation and Setup

The *SignaLink USB* comes with well-written and comprehensive *Installation & Operation* instructions. We won't repeat them here, but will briefly outline the procedure and provide information that may help avoid a few potential pitfalls.

Installing the Jumpers

The first thing that one must do is open the case and install jumper wires into the socket used to route various signals to the transceiver interface connector pins. This is actually quite simple to do, with TigerTronics providing the required jumper configuration for just about any radio that you might have. On the chance that your rig is not covered, they have a section in the manual that helps you determine how the jumpers need to be configured to support any radio you might have.

Installing the Window's Audio Codec Driver

After configuring the jumper wires, you connect the *SignaLink* to your computer so that it can automatically detect, locate and install the required drivers for supporting its USB Audio Codec. It turns out that most windows systems already have the driver; it just needs to be enabled. For Win98 installation, the manual says you need the original Win98 Installation CD, however my older Compaq Presario did not come with one, which gave me some concern. I decided to proceed with the installation anyway, and Win98 found the necessary drivers on my system without having to insert an Installation CD.

Sound Card Selection

Once the *SignaLink USB* has been detected and the drivers installed, your system "thinks" it has two sound cards installed. This brings us to the next point of the setup process; you need to make sure that for normal Windows system and application purposes, that the original sound card that came with your computer is selected. Likewise, for purposes of your digital communication program, in our case *DigiPan*, that the USB connected Audio Codec internal to the *SignaLink* is selected for it. If the system's sound card is not properly selected, your system will be unable to produce any sounds—a sure sign that you need to configure the setting.

For the following steps, make sure that the *SignaLink USB* is plugged into your system's USB port.

The sound card selection instructions that come with the unit can be followed in most instances. It's a matter of going to the **Control Panel** and configuring the default **Sound Playback** and **Sound Recording** devices found under the **Audio** tab of the **Sounds and Audio Devices** properties setup window. However, if you have Win98, you will instead use the same method to access the **MultiMedia Properties** window.

When you have located the correct setup window, use the pull-down selection box to set both the **Recording** and **Playback** devices to the sound card that came with your computer. The list should be short, probably only two or three entries. The name for the system's sound card will vary, on my Win98 system it was called **Via Audio (WAVE),** on my WinXP system it was called **SigmaTel Audio**.

Note: If using Vista, right click on the speaker icon in the task bar and select: **Sounds**, then the **Playback** tab. Select your computer's sound card from the list and set it back to the default sound card for playback. Next select the **Recording** tab, and select your computer's microphone as the default card for recording.

Once you have accomplished the sound card selection setup, your computer is then configured to send all system sounds to your computer's sound card and speakers. Back like it was before installing the USB Audio Codec drivers.

Now we still have to "tell" *DigiPan* which sound card to use. Click on DigiPan's **Configure** button and select **Sound card** from the pull-down menu. On the **Sound card** window make the following selections:

- **Type** **Computer soundcard**
- **Input** **USB Audio Device** (or some similar name)
- **Output** **USB Audio Device**

Setting the RX and TX Drive Levels

The *SignaLink USB* has front-panel RX and TX drive level control knobs, and since adjustment of the receive audio (waterfall drive level) is done completely within the *SignaLink* hardware; the recording level does not need to be set on the PC. On the other hand, even though there is a TX control knob on the *SignaLink*, the computer system's Volume Level still needs to be set as a master overall range setting. Once set, you can generally forget about it and use the *SignaLink's* TX drive level control to make any needed adjustments. The system's Volume Level setting interacts with the *SignaLink's* internal VOX level / gain setting; you may need to find an optimal compromise setting. Follow the procedures in the user manual.

Note: If using Vista, refer to page27 for instructions on setting the PC system's TX drive level volume control range.

Using the SignaLink with DigiPan

Once you have successfully installed the *SignaLink* as described in the *SignaLink's* user manual, you are ready to configure *DigiPan* for operation by continuing with Chapter 3, "Setting up DigiPan Software" in this guide. As you go through *DigiPan's* setup instructions and operating procedures, skip over the sections having to do with setting the: sound card type, serial port selection and adjusting the waterfall and transmit drive levels, as they have already been accomplished once the *SignaLink* installation procedure has been completed.

An Alternative USB Sound Card Interface

West Mountain Radio has recently introduced the *RIGblaster Advantage* which connects to the PC via USB and has an internal sound card, similar to the *SignaLink*, thus providing many of the advantages described in this chapter.

Like the *SignaLink* the *RIGblaster Advantage* is designed to support a wide variety of different radios, but also has a variety of other features that can provide greater flexibility in configuring your station. At the time this was written, the list price of the *RIGblaster Advantage* on the West Mountain Radio web site was $199.95, including a kit of interface connectors and cables for interfacing to most radios. Additional information is available on their web site: **http://www.westmountainradio.com/**

Chapter 8: **Onward to other Digital Modes**

Now that you have your rig running PSK, there is almost no limit to other digital modes of operation that you can try. Our home-brew interface, the *Rascal GLX*, the *SignaLink* and most other sound card to radio interfaces support a wide variety of sound card driven digital modes of operation. A sound card interface is the gateway to a whole new world of amateur radio operation.

The hardware interface only provides the connection between the PC's sound card and the transceiver; it's application software that provides the capability for other modes of operation. While *DigiPan* lacks the capability to operate these modes, other programs will do a variety of them.

If you have a compatible software program, the following digital modes of operation are a representative sample of what can be supported:

• FSK/RTTY	Audio Analyzer	MMTTY
• CW	Hellschreiber	WeatherFax
• Packet	EchoLink	WeFAX
• Pactor-1	Feldhell	MT63
• SSTV	MMSSTV	APRS

Where to Find Other Digital Mode Programs

We used *DigiPan* because it is widely used, simple to operate, works well and is free. However, it lacks many features that you might find desirable. Many of the other digital modes communication programs come with a fairly extensive set of features: improved logging, extensive log statistics and sorting, call lookup, CAT rig control, rotor control, mapping, multiple operating modes and lots more. Keep in mind, that all those features add complexity; *DigiPan* being more of a single purpose application is quite simple in comparison.

If you purchased a commercial interface, one of the first places to look is on the CD that came with your unit; it will generally be loaded with a variety of digital mode information and programs.

Another great resource is the Internet. Here are two web pages that provide links to a variety of digital modes programs:

AC6V's web page has an extensive list of PSK31 related web sites and programs at: **http://www.ac6v.com/software.htm#DIGITAL**

WM2U maintains a list of most of the communication programs on his web page at: **http://www.qsl.net/wm2u/psk31.html**

A few popular Multimode programs

HamScope **http://www.qsl.net/hamscope/**

HamScope supports PSK-31(BPSK & QPSK), MFSK16, RTTY, CW, ASCII and Packet. It also provides a radio control interface for many ICOM, TenTec, Kenwood and Yaesu radios. *HamScope* is freeware and runs under most versions of Windows.

HamRadio Deluxe **http://www.ham-radio-deluxe.com/**

HamRadio Deluxe along with its companion program, *Digital Master 780* supports PSK-31 (BPSK & QPSK), MT63, RTTY, Hellschreiber, CW, OLIVIA and THROB. It is designed to control transceiver's through the radio control interface available on many ICOM, TenTec, Elecraft, FlexRadio, Kenwood and Yaesu radios. Version 5.24 of *HamRadio Deluxe* is available as freeware and runs under most versions of Windows, the latest versions of the program may be purchased.

MixW **http://www.mixw.net/**

A very comprehensive Multimode shareware program. This thing is feature packed! *MixW* supports almost every mode imaginable, including CW, BPSK / BPSK63, QPSK, FSK31, RTTY, Packet (VHF/UHF), Pactor RX, AMTOR RX & FEC TX, MFSK/Graphics Color/BW, Throb, MT63, Hellschreiber, FAX, and SSTV. Standard and Contest Logging with CDROM Callbook lookup are also included. There are really too many features to list, refer to the above URL for more information. *MixW* runs on most versions of Windows and can be downloaded and tried free for 15-days, after that it will need to be registered at a cost of $50.

fldigi http://www.w1hkj.com

fldigi is a very popular freeware program, supporting PSK-31(BPSK & QPSK), MFSK16, RTTY, CW, DominoEX and many others. It comes with *flarq*, a companion program used to transfer data files. Perhaps uniquely, it also has French, Italian and Spanish translation files. *Fldigi / flarq* runs under Linux and most versions of Windows. A very good introductory / beginners guide can be found at **http://www.w1hkj.com/beginners.html**

One of the problems, and it's a good one to have, in selecting a digital modes program, is that there are so many to choose from. Another problem is that it's often difficult to know whether you are going to like a program until after you have down loaded, installed and used it.

This can be a fun but time consuming process. Talk to your friends, see what they like, perhaps they will even let you sit down and use theirs for a while. Finding a program you like is similar to finding a transceiver you like, once the basics are met, it's really a matter of personal preference. Good Hunting!

Chapter 9: **Other PC to Radio Interfaces**

In this guide we used the BuxComm *Rascal GLX* and the TigerTronics *SignaLink* interfaces as examples of different approaches to rig interfacing. The fact that they are discussed in detail in no way implies that we think they are the best solutions available. Like most things, the PC-to-Radio interface solution is a highly personal choice, depending upon your individual requirements and budget.

Sound Card to Radio Interfaces

The interfaces mentioned below cover a wide range of products, ranging from relatively simple interfaces to rather complex interfaces that attempt to do everything.

Rig Blasters by West Mountain Radio
http://www.westmountainradio.com/

West Mountain Radio has several models in their very popular *RIGblaster* series of interfaces, some of which are USB connected. The *RIGblasters* can come very sophisticated like the *RIGblaster Pro* or the *RIGBlaster Plus II,* which have a very broad and interesting set of features. The more recent *RIGblaster Advantage* connects to the PC via USB and has an internal sound card, similar to the *SignaLink* described in Chapter 7, but with more features. At the low-end, the *RIGblaster Nomic* and *RIGblaster Plug & Play* are lower cost, barebones simple interfaces to operate. Prices range from about $70 to $250. These products are available from most ham radio equipment distributors.

Sound Card-to-Rig Interfaces by MFJ
http://www.mfjenterprises.com

MFJ manufactures several different rig interfaces, priced from about $60 to $140, depending upon the features supported. MFJ interfaces are available from many ham radio equipment distributors.

Appendix: **Radio Setup Guides**

Nifty Ham Accessories produces laminated guides for almost all Kenwood, Icom and Yaesu radios sold since the year 2000.

Nifty set up and programming guides are available from most ham radio retailers and also directly from Nifty Ham Accessories.

Nifty! Quick Reference Guides Available for all recent Model Kenwood, Icom and Yaesu Transceivers

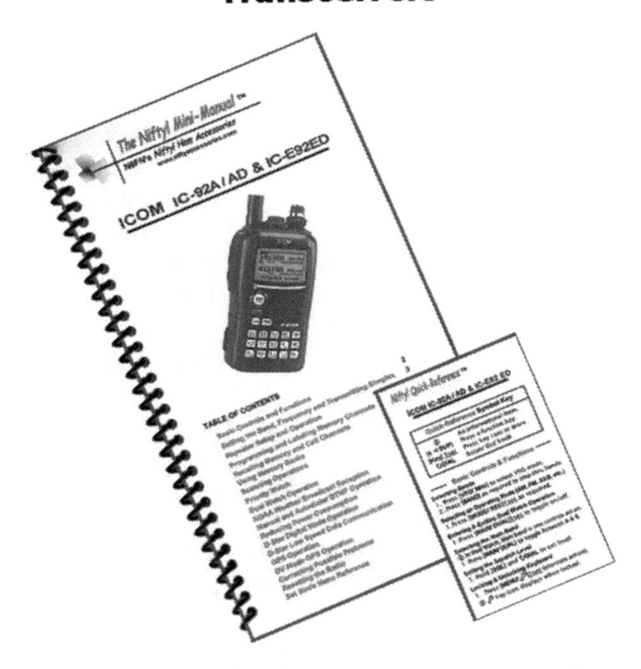

Condensed and simplified step-by-step operating instructions for all menus and modes of operation.

KENWOOD®

Transceiver Reference Guides

TH-K20AT / K40AT	TM-271A
TH-F6A / F7E	TS-480HX /SAT
TH-D7A(G)	TS-570 D/S(G)
TH-G71A	TS-590S
TH-D72A	TM-D700A
TM-V7A	TM-G707A
TH-22 / 42AT	TM-D710A
TS-50S	TM-V708A
TM-V71A	TS-870S
TM-261 / 461A	TS-2000 / 2000(X)

www.niftyaccessories.com

Icom

Transceiver Reference Guides

IC-T7H	IC-80AD	ID-880H
IC-Q7A	IC-91A / AD	IC-746PRO
IC-P7A	IC-92A / AD	IC-756PRO
IC-V8	IC-208H	IC-756PRO(II)
IC-R20	IC-910H	IC-756PRO(III)
IC-T22 / 42A	IC-2100H	IC-910H
IC-W32A	IC-2200H	IC-7000
IC-T70A	IC-2300H	IC-7200
IC-T81A	IC-2720H	IC-7410
IC-T90A	IC-2820H	IC-7600
IC-V80	IC-703 (P)	IC-7700
IC-V85	IC-706MKII	IC-7800
IC-T90A	IC-706MKII(G)	IC-9100
ID-31A / E	IC-718	IC-V8000

YAESU

Transceiver Reference Guides

VX-1R	FT-60R	FT-1000MP MKV
VX-2R	FT-90R	FT-1000MP Field
VX-3R	FT-100D	FT-2000 / FT-2000D
VX-5R	FT-250R	FT-2600M
VX-6R	FT-270R	FT-2800M
VX-7R	FT-450	FT-2900R
VX-8R	FT-817 / 817ND	FT-7100M
VX-8DR	FT-847	FT-7800R
VX-8GR	FT-857 / 857D	FT-7900R
VX-120	FT-897 / 897D	FT-8800R
VX-127	FT-920	FT-8900R
VX-150	FT-950	FTdx3000
VX-170 / VX-177	FT-1500M	FTdx5000
VR-500	FT-1802M	FTM-10R / 10SR
FT-50R	FT-1900R	FTM-350R

www.niftyaccessories.com